THE STIRLING
WATER-TUBE BOILER

THE BABCOCK & WILCOX CO.
NEW YORK

Copyright, 1912, by The Babcock & Wilcox Company

WORKS OF THE BABCOCK & WILCOX COMPANY, AT BAYONNE, NEW JERSEY

THE BABCOCK & WILCOX CO.

85 LIBERTY STREET, NEW YORK, U. S. A.

Works

BAYONNE . NEW JERSEY
BARBERTON . . . OHIO

Directors

E. H. WELLS, *President*	J. E. EUSTIS, *Secretary*
W. D. HOXIE, *1st Vice-President*	F. G. BOURNE
E. R. STETTINIUS, *2d Vice-President*	O. C. BARBER
J. G. WARD, *Treasurer*	C. A. KNIGHT

Branch Offices

ATLANTA	CANDLER BUILDING
BOSTON	35 FEDERAL STREET
CHICAGO	MARQUETTE BUILDING
CINCINNATI	TRACTION BUILDING
CLEVELAND	NEW ENGLAND BUILDING
DENVER	435 SEVENTEENTH STREET
HAVANA, CUBA	104 CALLE DE AGUIAR
HOUSTON	BRAZOS HOTEL BUILDING
LOS ANGELES	321 AMERICAN BANK BUILDING
NEW ORLEANS	SHUBERT ARCADE
PHILADELPHIA	NORTH AMERICAN BUILDING
PITTSBURGH	FARMERS' DEPOSIT BANK BUILDING
PORTLAND, ORE.	WELLS-FARGO BUILDING
SALT LAKE CITY	313 ATLAS BLOCK
SAN FRANCISCO	99 FIRST STREET
SEATTLE	MUTUAL LIFE BUILDING
TUCSON, ARIZ.	SANTA RITA HOTEL BUILDING

Export Department, New York: Alberto de Verastigui, Director

TELEGRAPHIC ADDRESS: FOR NEW YORK, "*GLOVEBOXES*";
FOR HAVANA, "*BABCOCK*"

WORKS OF THE BABCOCK & WILCOX COMPANY, AT BARBERTON, OHIO

FOREWORD

THE selection of steam boilers is a matter which is worthy of the most careful thought and attention. While some purchasers of boilers give this important subject proper consideration, there are many who accept as entirely satisfactory those boilers with which they have had experience, or who buy the cheapest after little or no investigation. Purchasers of the latter class often carefully consider the comparative economy and reliability of engines and auxiliaries, so that in some steam plants we find refinements of economy and convenience in the engines and auxiliaries, while the boilers may be wasteful in operation and deficient in the essentials of simplicity, economy and adaptability to the service for which they are used. While refinements in engine economies add materially to the first cost of the plant, the best boiler may frequently be purchased at a cost comparatively little above that of an inferior one, and will effect a greater saving for a given increase in cost than could possibly be obtained by installing a more efficient engine.

The special requirements of each individual case should be carefully considered before determining the general type of the boiler needed. After the general type has been decided upon, the most important features to be considered are safety, efficiency, durability and accessibility. Of almost equal importance are the experience, skill, financial responsibility and reputation of the manufacturers.

The method of supporting the boiler should provide for the free expansion of the component parts under changes of temperature without introducing unequal strains. The circulation should be such as to keep all parts at practically the same temperature. All parts of the boiler should be so accessible for inspection as to prohibit the possibility of concealed corrosion.

Baffles should be so designed that they cannot be easily damaged or displaced in service or by gas explosions, and should be so accessible that if a defect be discovered it may be readily repaired without removing tubes or any part of the setting.

The arrangement of tubes should be such that any tube may be removed and replaced without disturbing any other tube. The spacing should allow the free passage of the gases around each tube so as to avoid the possibility of the gases being throttled or of the spaces becoming clogged by deposits of soot. Sufficient steam and water capacity should be provided to insure dry steam under widely varying load conditions.

Riveted seams should not be placed in the path of the hottest gases. There should be no possibility of steam or air pockets at points exposed to intense heat. Tubes that may become overheated in case of low water should not be used as stays. Staybolts are wholly objectionable. The sole useful purpose which a staybolt serves in a stationary boiler is to make it possible to use a cheap form of construction. Check valves or other delicate mechanical devices should not be used in the interior of a boiler.

BLACKSTONE HOTEL, CHICAGO, ILL., OPERATING 1012 HORSE-POWER OF STIRLING BOILERS

Large flat surfaces, stayed or unstayed, are among the most dangerous and otherwise objectionable features in boiler construction and should not be used in any boiler carrying the high pressures now common in modern steam plants.

Another objection to flat stayed surfaces is that in vertical water-tube boilers the flat surfaces are usually so located as to form a convenient lodging place for flue dust. The flue dust fuses into a hard mass which is difficult to remove,

and which, on account of its non-conductivity, destroys the effectiveness of that portion of the heating surface. Further, such an accumulation increases the chance of corrosion without the possibility of detection. The stays collect scale and mud and increase the difficulty and expense of cleaning.

The life of a good boiler is variable, depending upon the attention it receives, but a modern, properly designed water-tube boiler should be capable of standing the test of a long period of years without material reduction in its margin of safety and economy.

INTERNATIONAL SMELTING AND REFINING COMPANY, TOOELE, UTAH, OPERATING 4780 HORSE-POWER OF STIRLING BOILERS

HISTORY OF THE STIRLING WATER-TUBE BOILER

THE Stirling boiler was first manufactured commercially by The International Boiler Company, Limited, of New York, in 1889. The first boilers built consisted of two upper drums and a lower drum, being crudely constructed and having little or no attention given, either in construction or erection, to those minor details which in the aggregate make for the success of a boiler. Crude though the construction of these boilers was, they demonstrated that the design was such as to give great possibilities for development. With this fact established, The Stirling Boiler Company was formed and purchased the interests of The International Boiler Company, Limited, in 1890.

The construction of the boiler was elaborated, though the principle remained the same. A third upper drum was added, the plan of setting modified, and such improvements made as would naturally result from a systematic effort to produce a safe, durable and economical steam generator for the varying conditions which are met in steam practice.

In 1905 The Stirling Boiler Company acquired other interests and became The Stirling Consolidated Boiler Company. In 1907 the manufacturing plant and business of The Stirling Consolidated Boiler Company were acquired by The Babcock & Wilcox Company.

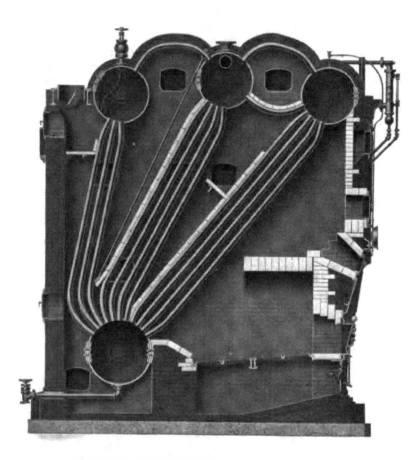

SECTIONAL ELEVATION, CLASS "A" STIRLING BOILER

GENERAL DESCRIPTION

THE Stirling boiler is built in a number of different designs, known as classes, to meet varying conditions of floor space and head room. All classes are of the same general design varying in depth, height and in the number and length of the tubes. The boiler consists of three transverse steam-and-water drums, set parallel, and connected to a mud drum by water tubes, so curved as to enter the tube sheets radially. The steam space of the center drum is interconnected to both the front and rear drums by a row of curved steam circulating tubes and to the water space of the front drum by water circulating tubes, the number of these latter tubes depending upon the class of the boiler. The main steam outlet is placed on the top of the center drum. Two independent safety valves are also placed on the top of this drum, and to one drum head a water column is connected.

A feed pipe enters the top of the rear steam-and-water drum at the center and discharges into a removable trough, by which the feed water is distributed over a relatively large width of the drum.

A blow-off connection, or connections, depending upon the size of the boiler is placed at the bottom of the mud drum and is extended through a sleeve in the rear or side wall, just outside of which the blow-off valve is located.

The pressure parts of the boiler are supported on saddles under each steam-and-water drum by a rectangular structure of rolled steel sections entirely independent of the brickwork.

FORGED STEEL DRUM HEAD
WITH MANHOLE PLATE
IN POSITION

DRUM CONSTRUCTION — Each drum is made of a single tube sheet riveted by properly proportioned lap or butt and strap longitudinal seams to a drum sheet. The drum heads are of forged steel, one head in each drum being provided with a manhole fitted with a forged steel manhole plate and guards.

TUBE SPACING — Sufficient space is left between the tubes to permit a free passage of the gases. The tubes are so spaced that any tube can be removed and replaced without disturbing any other tube or the brickwork. After a tube has been removed it is passed out through one of the doors built into the setting for that purpose.

BAFFLES — The baffle brick are plain fire brick tiles resting against the rear tubes of the first and second banks, reaching in the first instance from the mud drum nearly to the top of the first bank, and in the second instance from the center steam-and-water drum nearly to the bottom of the second bank. A shelf

PARTIAL FRONT ELEVATION AND SECTIONAL ELEVATION THROUGH FURNACE AND FRONT STEAM DRUM

placed near the top of the front baffle deflects the gases into the second bank of tubes. A second shelf is placed near the top of the rear bank of tubes and deflects the gases in their passage upward through this rear bank into the tubes, thus preventing by-passing between the tubes and the rear wall of the boiler setting. A covering of fire brick resting on the water circulating tubes between the front and middle steam-and-water drums prevents the gases passing above these tubes. The baffle walls guide the gases up the front bank of tubes, down the middle bank and up the rear bank, bringing them into intimate contact with all of the heating surfaces. The baffle openings between the banks are so designed that there will be a proper distribution of the products of combustion with a minimum amount of throttling action, and can be readily adjusted to suit fuel conditions.

FORGED STEEL DRUM HEAD
INTERIOR

DAMPER BOX — A damper box equipped with a swinging damper is placed either on the top of the boiler at the rear of the setting or in the rear wall. In the first instance it rests on special supports carried on the boiler supporting framework. With such an arrangement it may either in turn support an overhead stack and breeching or it may be connected to an overhead flue. In the second case, the damper frame is built into the rear wall and is adaptable for any method of rear flue connection.

EXPANSION — The mud drum is suspended from all of the steam-and-water drums by the water tubes, swinging entirely free of the setting brickwork. Air leakage around the ends of this drum is prevented by soft asbestos packing between it and the brickwork. This construction, together with the curvature of the tubes which is necessary in order that they may enter the tube sheets radially, gives ample and efficient provision for expansion and contraction. The design ensures thorough equalization and proper distribution of all strains incident to the service of steam generation.

BRICKWORK — The setting of the Stirling boiler is simple, being rectangular in outline. No special shapes of bricks not to be found on the open market are required for the setting, and the work may be done by any brick-mason familiar with furnace brickwork and who can read drawings. The arrangement of arch skewbacks is of such a nature that a complete furnace lining may be installed without in any way disturbing the boiler arch. All masonry repairs to the brickwork may be done without disturbing the pressure parts of the boiler or its connections.

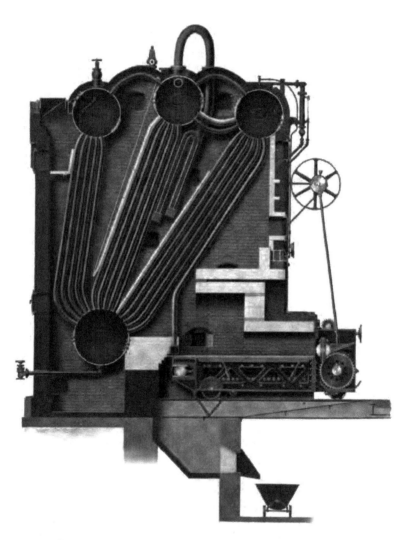

SECTIONAL ELEVATION, CLASS "S" STIRLING BOILER WITH BABCOCK & WILCOX
SUPERHEATER AND BAYONNE CHAIN GRATE STOKER

FURNACE — The design of the Stirling furnace possesses many distinctive advantages. By referring to the illustrations it will be seen that a fire-brick arch is sprung over the grates and immediately in front of the first bank of tubes. The large triangular space between the boiler front, the tubes and the mud drum forms a fire-brick combustion chamber in which it is possible to install a sufficient amount of grate surface to meet the requirements of the lowest grades of fuel. The arch, acting in a manner similar to the roof of a reverberatory furnace, heats any air admitted over the fuel bed, and the gases distilled from the fuel are ignited by the heat radiated from the arch. It ensures a proper distribution of the gases to the front bank of tubes and prevents the cooling of the boiler by any inrush of cold air when the furnace doors are opened. The gases do not come into contact with the comparatively cool surfaces of the tubes until after they have passed out of the fire-brick combustion chamber in which they have had ample space and time to be properly burned. The furnace is readily adaptable to the fuel available, whether solid, liquid or gaseous.

PORTION OF 19400 HORSE-POWER INSTALLATION OF STIRLING BOILERS
FOR THE BETHLEHEM STEEL CO., SO. BETHLEHEM, PA.

FRONT — The front is of ornamental design, substantially made of cast iron and steel, and is built up in sections and bolted together. The joints are so placed as to permit the application of any stoker, oil or gas burners. The joints provide for all expansion and thus prevent warping and cracking.

ACCESS AND CLEANING DOORS — Doors for cleaning the heating surfaces and for access to the interior of the brick setting are provided in the front, side and rear walls in sufficient number to allow all parts to be thoroughly cleaned by the means of a steam lance and to make the exterior of the heating surfaces easily accessible for inspection. All cleaning doors seat tight against asbestos packing hammered into a groove in the face of the door frame, in this manner preventing the leakage of air into the setting.

ACCESS AND CLEANING DOOR

A large circular door in the setting gives access to the manhole end of the mud drum. This door is also asbestos packed to prevent air leakage.

INTERIOR CLEANING — Removing the manhole plates from the four drums gives easy access to the interior of all heating surfaces for examination, cleaning and repairs.

MUD DRUM ACCESS DOOR

Any scale which may have formed on the interior surfaces of the tubes may be removed by a turbine cleaner of any of the many designs on the market. The hose to which the cleaner is attached is passed into the drum, the operator running the cleaner through the tubes by means of the hose.

FITTINGS — The boiler accessories consist of the following:

Feed water connections and valves attached to the rear steam and water drum.

Blow-off connections and valves connected to the mud drum.

Safety valves placed on the center steam-and-water drum.

A water column connected to the center steam-and-water drum and placed in a position visible from any point forward of the boiler front.

A steam gauge attached to the boiler front.

All of these fittings are substantially made and are of designs which by their successful service for many years have become standard with The Babcock & Wilcox Company.

OPERATION — The path of the gases from the furnace has already been indicated. The water which as stated is fed into the rear steam-and-water drum, passes downward through the rear bank of tubes to the mud drum, thence upward through the front bank of tubes to the forward steam-and-water drum.

SMITH BREWING COMPANY, YOUNGSTOWN, OHIO, OPERATING 863 HORSE-POWER OF STIRLING BOILERS

The steam formed during the passage upward through the front bank of tubes becomes separated from the water in the front drum and passes through the upper row of cross tubes or steam circulators into the center steam-and-water drum, from which point it passes through the dry pipe into the steam main. The water from the front drum passes through the lower or water circulating tubes into the middle drum and thence downward through the middle bank of tubes to the mud drum, from which it is again drawn up the front bank to retrace its course. The steam generated in the rear bank of tubes passes through the rear steam circulators to the center steam-and-water drum.

The great water storage capacity of the four drums and the tubes, together with the large disengaging surface of the three steam-and-water drums, and the arrangement by which the greatest steam space is in the steam-and-water drum from which the steam is taken, ensures the production of dry steam under varying loads and irregular firing conditions.

BAD FEED WATER — In its passage downward through the rear bank of tubes the feed water is heated to such an extent that much of the scale forming matter is precipitated and gathers in this bank and in the mud drum. Here it is protected from high temperatures and can be washed and blown down as frequently as the case demands. As the circulation is comparatively slow in the rear bank of tubes a large percentage of matter held in suspension is deposited in the mud drum before reaching that portion of the heating surface subjected to intense heat.

MATERIALS AND WORKMANSHIP — The details of the Stirling boiler have been developed after years of most careful observation on the part of competent engineers. Materials entering into their construction are the best obtainable for the special purpose for which they are used and are subjected to rigid inspection and tests. All pressure parts have a factor of safety of at least 5.

The boilers are manufactured by the most modern shop equipment and appliances in the hands of an old and well trained organization of skilled mechanics under the direct supervision of experienced engineers.

INDIANAPOLIS LIGHT AND HEAT COMPANY, INDIANAPOLIS, IND., OPERATING 8800 HORSE-POWER OF STIRLING BOILERS

CIRCULATION

A WELL designed water-tube boiler should possess to a high degree of perfection the important feature of definite and positive circulation.

The effect of different degrees of expansion in different parts of the structure, so destructive to cylindrical and tubular boilers, is eliminated in a properly designed water-tube boiler. The difference in expansion of the various parts of a boiler is dependent upon the difference in the temperature of those parts; consequently the greater the uniformity in temperature of the water the less will be the difference in expansion between the different parts of the boiler, as the temperature of the pressure parts (when the material is not too thick) must be practically the temperature of the contained water.

Rapid circulation insures uniformity of temperature of the water and of the pressure parts and thereby prevents unequal expansion and contraction with the subsequent destructive strains.

In the Stirling boiler the rapid circulation carries the steam bubbles with the current of water to the disengaging surface and the steam space, and thus prevents the formation of steam pockets and the consequent overheating and burning of tubes at points where the greatest heat is applied.

U-TUBE ILLUSTRATING CIRCULATION IN A PROPERLY DESIGNED WATER-TUBE BOILER

The theory of circulation, as described by Geo. H. Babcock, presents the matter in a clear and most satisfactory manner. His discussion of the subject in "Steam" is too well known to require repetition here. The circulation is illustrated by applying the flame of a lamp to one leg of a U-tube, suspended from the bottom of a vessel filled with water, the heat from the flame setting up a uniform circulation, as indicated in the illustration. Mr. Babcock states: "This U-tube is the representation of the true method of circulation within a water-tube boiler properly constructed."

The sectional views of the Stirling water-tube boiler on the preceding pages, will indicate that the design is such as to fully meet the requirements for uniform circulation as illustrated by the U-tube and flame. The front bank of tubes, subjected to the most intense heat, represents the leg of the U-tube to which the flame is applied. The uniform circulation up the front bank of tubes, through the water circulating tubes to the center steam-and-water drum and down the center bank of tubes to the mud drum, together with the downward circulation through the rear bank of tubes to replace water evaporated into steam, ensures a complete and clearly defined circulation throughout the entire boiler.

GUANICA CENTRAL, GUANICA, PORTO RICO, OPERATING 5970 HORSE-POWER OF STIRLING BOILERS

THE STIRLING BOILER IN SERVICE

STIRLING boilers have been in operation since 1890, and their performance since that time has clearly demonstrated their right to all of the claims of excellence which have been made for them.

The ease with which the Stirling boiler may be cleaned, its efficient and substantial baffling and its flexibility under varying load conditions, have caused it to be adopted extensively in plants representing practically every industry throughout the world. Over 3,000,000 horse-power of Stirling boilers are in use in electric light and power plants, street railway power stations, coal mining plants, blast furnaces, rolling mills, smelting and refining plants, heating and lighting plants in educational institutions, sugar mills, breweries, cotton mills, lumber mills, ice plants, oil refineries, and their allied industries.

The Stirling boiler has proved entirely successful in the use of anthracite and bituminous coals with both hand and stoker firing, lignite from the various lignite fields, oil fuel, wood and saw mill refuse, green bagasse, tan bark, blast furnace, coke oven and natural gas, and waste heat from brick kilns, cement kilns and smelting furnaces.

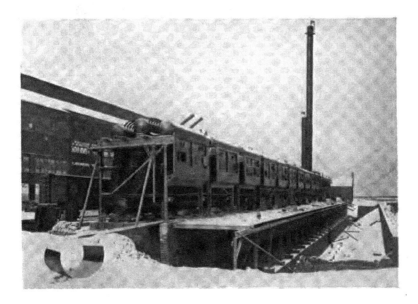

OHIO WORKS OF THE CARNEGIE STEEL COMPANY, YOUNGSTOWN, OHIO, OPERATING 20,500 HORSE-POWER OF STIRLING BOILERS

CARE AND MANAGEMENT OF THE STIRLING BOILER

BEFORE placing a new boiler in service a careful and thorough examination should be made of the pressure parts and the setting. The latter should be inspected to see that the baffle openings and the distance from the arch to the tubes are as called for by the particular drawings for the installation in question; that the joints of the baffle tile are directly behind the tubes; that the mud drum and blow-off pipe are free to expand without interference with the setting walls; and that all brick and mortar are cleaned from the setting and pressure parts. Tie rods should be set up snug and then slacked slightly until the setting has been thoroughly warmed after the first firing. Internally the boiler should be examined to insure the absence of dirt, waste, oil and tools.

If there is oil or paint in the boiler, one peck of soda ash should be placed in each upper drum, the boiler filled to its normal level with water and a slow fire started. After twelve hours the fire should be allowed to die out, the boiler cooled slowly, then opened and washed out thoroughly. This will remove all oil and grease from the interior of the boiler and prevent foaming when it is placed in service.

The water column piping should be examined and known to be free and clear, and the water level as indicated by the gauge glass should be checked by opening the gauge cocks.

Firing the boiler with green walls will invariably crack the setting brickwork unless this be dried properly. To start this drying process, as soon as the brickwork is completed the damper and ash pit doors should be blocked open to maintain a circulation of air through the setting. Whenever possible, this should be done for several days before firing. When ready for firing, wood should be used for a light fire, gradually building it up until the walls are thoroughly warmed. Coal should then be fired and the boiler placed in regular service.

A boiler should not be cut into the line with other boilers until the pressure is within a few pounds of that in the steam main. The boiler stop valve should be opened very slowly until it is opened fully. Care must be taken to see that the arrangement of piping is such that there will be no possibility of water collecting in any pocket between a boiler and the main, from which it can be carried over into the steam line when the boiler is cut in.

In regular operation the safety valve and the steam gauge should be checked daily. The steam pressure should be raised sufficiently to cause the safety valves to blow, at which time the steam gauge should indicate the pressure for which the safety valves are known to be set. If it does not, one is in error and the gauge should at once be compared with one of known accuracy and any discrepancy rectified. The water column should be blown down thoroughly at least

WORCESTER ELECTRIC LIGHT COMPANY, WORCESTER, MASS., OPERATING 3600 HORSE-POWER OF STIRLING BOILERS

once on each shift and the height of the water as shown by the gauge glass checked by opening the gauge cocks at the side of the column. The bottom blow-off valves should be kept tight and opened at least once daily to blow from the mud drum any sediment which may have collected from concentration of the boiler feed water. The amount of blowing necessary will depend upon the character of the feed water used.

In case of low water, resulting either from carelessness or unforeseen conditions of operating, the essential object to be attained is to extinguish the fire in the quickest possible manner. Ordinary practice has been to cover the fires with wet ashes, dirt or fresh fuel. Under certain conditions it is feasible to put out the fires with a heavy stream of water from a hose and

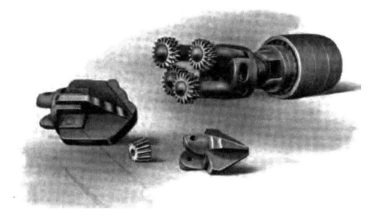

FIG. 1

this method, where practicable, should be followed. The boiler should be cut out of the line and a thorough inspection made to ascertain what damage, if any, has been done before it is again placed in operation.

The efficiency and capacity of a boiler depend to an extent very much greater than is ordinarily appreciated upon its cleanliness internally and externally, and systematic cleaning should be included as a regular feature in the operation of any steam plant.

The outer surfaces of the tubes should be blown free from soot with a steam lance at regular intervals, the frequency of such cleaning periods depending upon the class of fuel burned. Internally the tubes should be kept clean from scale and sludge which will accumulate due to the concentration of solids present in practically any boiler feed water. This internal cleaning can best be accomplished by the use of an air or water driven turbine, the cutter heads of which

GARFIELD SMELTING COMPANY, GARFIELD, UTAH, OPERATING 7000 HORSE-POWER OF STIRLING BOILERS

may be changed to handle varying thicknesses of scale. Figure 1 shows a turbine cleaner together with several cutting heads which has been found to give satisfactory results.

When scale has been allowed to accumulate to an excessive thickness the work of removing it is difficult. Where the scale is of a sulphate formation its removal may be made easier by filling the boiler with water in which there has been placed a bucketful of soda ash to each drum, starting a slow fire and allowing the water to boil for twenty-four hours without allowing any pressure on the boiler. It should then be cooled slowly, drained, and the turbine cleaner used immediately as the action of the air tends to harden the scale. While the use of a boiler compound in feed water is permissable with a view to preventing the formation of scale, such an agent should not be introduced into the boiler while it is in operation with a view to softening or loosening any scale that may already be present in the boiler.

Aside from the aspect of efficiency and capacity, a clean interior of boiler heating surfaces insures protection from burning. In the absence of a blow-pipe action of the flames, it is impossible to burn a metal surface when water is in intimate contact with that surface. Any formation of scale on the interior surfaces of a boiler will keep the water from those surfaces and increases their tendency to burn. Particles of loose scale which may have become detached will lodge at certain points in the tubes and act at such points in the same manner as a continuous coating of scale except that the tendency to burn is localized. If oil is allowed to enter the boiler with the feed water, its action will be the same as that of scale in keeping the water from the metal of the tubes, in this way increasing their liability to burn.

It has been proven beyond doubt that a very large percentage of tube losses is due to the presence of scale which, in many instances has been so thin as to be considered of no moment, and the importance of maintaining the interior of boiler heating surfaces in a clean condition cannot be emphasized too strongly.

If pitting or corrosion is noted, the parts affected should be carefully cleaned and painted with white zinc. The cause of such action should be determined immediately and steps taken to see that a proper remedy is applied.

When making an internal inspection of a boiler or when cleaning the interior of the heating surfaces, great care must be taken to guard against the possibility of steam entering the boiler in question from any other boilers on the line through open blow-off valves or through the careless opening of the boiler stop valve. Bad cases of scalding have resulted from neglect of this precaution.

Boilers should be taken out of service at regular intervals for cleaning and repairs. When this is done, the boiler should be allowed to cool slowly and when possible allowed to stand twelve hours after the fires are drawn before opening. The cooling process should not be hastened by causing cold air to rush through the setting as this will cause difficulties with the setting brickwork. While the

STIRLING BOILERS IN COURSE OF ERECTION AT THE UNIVERSITY OF CINCINNATI, CINCINNATI, OHIO

boiler is off for cleaning, a careful examination should be made of its condition, both internally and externally, and all leaks of steam, water and air through the setting should be stopped promptly.

If a boiler is to remain idle for some time it is liable to deteriorate much faster than when in service. If the period for which it is to be laid off is not to exceed three months it may be filled with water while out of service. The boiler should be thoroughly cleaned internally and externally, all soot and ashes being removed from the setting and any accumulation of scale removed from the interior surfaces. It should then be filled with water to which about four buckets of soda ash has been added, a very light fire started to drive the air from the water, the fire then allowed to die out and the boiler pumped full.

If the boiler is to be out of service for more than three months, it should be emptied, cleaned and thoroughly dried. A tray of quicklime should be placed in each drum, the boiler closed up, the grates covered and a quantity of quicklime placed on these. Special care must be taken to prevent air, steam or water leaks into the setting or onto the pressure parts, to obviate danger of corrosion.

4800 HORSE-POWER INSTALLATION OF STIRLING BOILERS FOR THE MARQUETTE CEMENT MANUFACTURING COMPANY, LA SALLE, ILL.

TESTS OF STIRLING BOILERS WITH VARIOUS FUELS

Plant			Harvard Medical School	Wilkesbarre Gas & Elec. Co.	Boston Elevated Ry. Co.
Location			Boston, Mass.	Wilk'barre, Pa.	Boston, Mass.
Boiler heating surface		sq. ft.	3188	2404	3500
Rated horse-power		H. P.	319	240	350
Type of furnace			Hand fired	Hand fired	Hand fired
Grate surface		sq. ft.	64	81	61
Fuel			Bit. run of mine	Anthracite rice	Bit. run of mine
Source of trade name			Pocahontas	Lehigh Valley	Pocahontas
Duration of test		hours	10	7.5	10
Steam pressure by gauge		lbs.	112	147.6	172
Temperature of feed water		°F.	75	34.6	167
Factor of evaporation			1.1824	1.2285	1.0945
Blast under grates		inches	....	1.16	.22
Draft in furnace		inches	.15	.12	.08
Draft at boiler damper		inches	.40	.25	.30
Temperature of escaping gases		°F.	479	582	625
Total water fed to boiler		lbs.	92288	76577	133822
Equivalent evaporation from and at 212°		lbs.	109121	94075	146468
Equivalent evaporation from and at 212° per hour		lbs.	10912	12543	14647
Equivalent evaporation from and at 212° per square foot of heating surface per hour		lbs.	3.42	5.22	4.18
Horse-power developed		H. P.	316.3	363.6	424.5
Per cent of rated horse-power developed		%	99.1	151.5	121.3
Total fuel fired		lbs.	10647	12449	14200
Per cent of moisture in fuel		%	3.12	4.08	4.09
Total dry fuel		lbs.	10315	11941	13619
Per cent of refuse by test		%	8.62	22.1	8.21
Total combustible		lbs.	9426	9302	12501
Dry fuel per square foot of grate surface per hour		lbs.	16.12	19.65	20.5
Flue gas analysis	CO_2	%	13.9	....	13.6
	O	%	5.2	....	6.2
	CO	%	.1	....	.4
Proximate analysis dry fuel	Volatile matter	%	....	6.13	17.73
	Fixed carbon	%	....	70.67	74.79
	Ash	%	....	23.20	7.48
B. T. U. per pound of dry fuel		B.T.U.	14381	11298	14637
Equivalent evaporation from and at 212° per pound of dry fuel		lbs.	10.58	7.88	10.75
Efficiency boiler and furnace		%	71.39	67.63	71.27

1505 HORSE-POWER INSTALLATION OF STIRLING BOILERS FOR THE UTAH IDAHO SUGAR COMPANY, ELSINOR, UTAH

TESTS OF STIRLING BOILERS WITH VARIOUS FUELS

		Gen. Elec. Co.	B. & W. Co.	Detroit Edison Co.
Plant		Gen. Elec. Co.	B. & W. Co.	Detroit Edison Co.
Location		Lynn, Mass.	Barberton, O.	Detroit, Mich.
Boiler heating surface	sq. ft.	10000*	11279	23650
Rated horse-power	H P.	1000	1128	2365
Type of furnace		Roney	B & W.chain gr.	Taylor
Grate surface	sq. ft	180	187	405
Fuel		Bitum. slack	Bitum slack	Bitum. slack
Source of trade name		New River	Pittsburgh	RedJckt.,W.Va
Duration of test	hours	24	8	26.5
Steam pressure by gauge	lbs.	171	132	210
Temperature of feed water	°F.	182	109	188
Degree of superheat	°F.	....	152	165.3
Factor of evaporation		1.0789	1.2352	1.1697
Blast under grates	inches	....	1.09	2.53
Draft in furnace	inches	.21	.16	.26
Draft at boiler damper	inches	.67	.97	.84
Temperature of escaping gases	°F.	667	624	651
Total water fed to boiler	lbs.	881348	499893	3907028
Equivalent evaporation from and at 212°	lbs.	950886	617468	4570051
Equivalent evaporation from and at 212° per hour	lbs.	39620	77184	172456
Equivalent evaporation from and at 212° per square foot of heating surface, per hour	lbs.	3.96	6.84	7.29
Horse-power developed	H. P.	1148	2237.2	4999
Per cent of rated horse-power developed	%	114.8	198.3	211.3
Total fuel fired	lbs.	92565	67292	424000
Per cent of moisture in fuel	%	5.23	3.50	1.9
Total dry fuel	lbs.	87724	64937	415944
Per cent of refuse by test	%	9.24	18.29	9.55
Total combustible	lbs.	79618	53060	376221
Dry fuel per square foot of grate surface, per hour	lbs.	20.31	43 41	38.75
Flue gas analysis { CO_2	%	11.3	11.2	15.45
O	%	7.5	8.3	3.86
CO	%	0.13	.0	0.17
Proximate analysis dry fuel { Volatile matter	%	20.28	31.35	33.48
Fixed carbon	%	75.84	52.71	60.58
Ash	%	3.88	15.94	5.94
B. T. U. per pound of dry fuel	B.T.U.	14833	12130	14061
Equivalent evaporation from and at 212° per pound of dry fuel	lbs.	10.84	9.51	10.99
Efficiency boiler and furnace	%	70.92	76.08	75.84

*Two boilers.

PORTION OF 3000 HORSE-POWER INSTALLATION OF STIRLING BOILERS FOR THE N. K. FAIRBANKS COMPANY, CHICAGO, ILL.

TESTS OF STIRLING BOILERS WITH VARIOUS FUELS

		Cincinnati W.W.	Muncie Electric Light Company	Poughkeepsie Lt., Ht. & Pr. Co.
Plant				
Location		Cincinnati, O.	Muncie, Ind.	P'hk'psie, N.Y.
Boiler heating surface	sq. ft.	4272*	5000	3060
Rated horse-power	H. P.	427	500	306
Type of furnace		Amer. stoker	Green chain gr.	Murphy
Grate surface	sq. ft.	84	95	....
Fuel		Bit. run of mine	Bitum. slack	Bit. run of mine
Source of trade name		Pittsburgh	Indiana	Willmore, Pa.
Duration of test	hours	144	8	8
Steam pressure by gauge	lbs.	156.4	157	150
Temperature of feed water	°F.	211.6	56	176
Degree of superheat	°F.	....	....	119
Factor of evaporation		1.0469	1.2073	1.1519
Blast under grates	inches	1.70	....	....
Draft in furnace	inches	.03	.39	.23
Draft at boiler damper	inches	.13	.90	.54
Temperature of escaping gases	°F.	407	411	560
Total water fed to boiler	lbs.	1372862	140871	118102
Equivalent evaporation from and at 212°	lbs.	1437249	170074	136042
Equivalent evaporation from and at 212° per hour	lbs.	9981	21259	17005
Equivalent evaporation from and at 212° per square foot of heating surface per hour	lbs.	2.34	4.25	5.56
Horse-power developed	H. P.	289.3	616.2	492.9
Per cent of rated horse-power developed	%	67.8	123.2	161.0
Total fuel fired	lbs.	142397	23829	12790
Per cent of moisture in fuel	%	1.77	10.30	5.42
Total dry fuel	lbs.	139877	21375	12097
Per cent of refuse by test	%	10.23	20.00	7.53
Total combustible	lbs.	125565	17100	11186
Dry fuel per square foot of grate surface per hour	lbs.	11.56	28.1	....
Flue gas analysis { CO_2	%	13.0	8.2	....
O	%	5.9	....	....
CO }	%	.0	....	....
Proximate analysis dry coal { Volatile matter	%	33.42	40.07	21.30
Fixed carbon	%	56.30	38.62	72.29
Ash }	%	10.28	21.31	6.41
B. T. U. per pound of dry fuel	B.T.U.	13332	10908	14797
Equivalent evaporation from and at 212° per pound of dry fuel	lbs.	10.28	7.96	11.25
Efficiency boiler and furnace	%	74.83	70.81	73.78

* Two boilers.

6000 HORSE-POWER INSTALLATION OF STIRLING BOILERS AT THE OPEN HEARTH PLANT OF THE REPUBLIC IRON AND STEEL COMPANY, YOUNGSTOWN, OHIO

TESTS OF STIRLING BOILERS WITH VARIOUS FUELS

		Pac.Lt.& Pr.Co	Southern Porto Rico Sugar Co	Ill. Steel Co.
Plant				
Location		Los Angeles, Cal.	Porto Rico	So. Chi., Ill.
Boiler heating surface	sq. ft.	3288	4685	3612
Rated horse-power	H. P.	329	468	361
Type of furnace		Oil	Exten. bagasse	Blast fur. gas
Grate surface	sq. ft.	. . . .	23.4	. . . .
Fuel		Whittier, Cal. Oil	Bagasse	Blast fur. gas
Duration of test	hours	10	6	1.85
Steam pressure by gauge	lbs.	156	84	136.5
Temperature of feed water	°F.	63.7	155.1	44
Factor of evaporation		1.1992	1.0957	1.2175
Blast under grates	inches	. . . .	.24	4.2‡
Draft in furnace	inches	.09	.13	.07
Draft at boiler damper	inches	.14	.35	.47
Temperature of escaping gases	°F.	454	539	743
Total water fed to boiler	lbs.	97722	95623	33650
Equivalent evaporation from and at 212°	lbs.	117188	104744	40969
Equivalent evaporation from and at 212° per hour	lbs.	11719	17462	22145
Equivalent evaporation from and at 212° per square foot of heating surface per hour	lbs.	3.56	3.73	6.13
Horse-power developed	H. P.	339.7	506.1	641.9
Per cent of rated horse-power developed	%	103.3	108.1	177.8
Total fuel fired	lbs.	7764	38952	. . . .
Per cent of moisture in fuel	%	1.06	44.70	. . . .
Total dry fuel	lbs.	7682	21540	. . . .
Dry fuel per square foot of grate surface per hour	lbs.	. . . .	153.4	. . . .
Flue gas analysis — CO$_2$	%	. . . .	13.2	20.1
Flue gas analysis — O	%	. . . .	6.9	4.0
Flue gas analysis — CO	%	. . . .	.1	.3
Ultimate analysis dry fuel — Carbon	%	86.02	44.12	Anal. ent'g gas
Ultimate analysis dry fuel — Hydrogen	%	11.52	6.30	CO$_2$ — 12.5
Ultimate analysis dry fuel — Oxygen	%	1.45	47.47	O — .0
Ultimate analysis dry fuel — Nitrogen	%	.25	0.41	CO — 25.8
Ultimate analysis dry fuel — Sulphur	%	.75	. . . .	H — 2.5
Ultimate analysis dry fuel — Ash	%	.01*	1.70	CH$_4$ — .2
B. T. U. per pound of dry fuel	B.T.U.	18677	4088	. . . .
Equivalent evaporation from and at 212° per pound of dry fuel	lbs.	15.25	4.86	. . . .
Efficiency of boiler and furnace	%	79.23	63.85†	. . . .

* Silt.
† Efficiency based on thermal value of bagasse as fired corrected for heat lost in evaporating and superheating its moisture. If correction extended to heat lost in burning H., efficiency would be 72.31%.
‡ Pressure of entering gas.

GENERAL VIEW OF FURNACE PLANT OF THE REPUBLIC IRON AND STEEL COMPANY, YOUNGSTOWN, OHIO

MERCHANTS HEAT AND LIGHT COMPANY, INDIANAPOLIS, IND., OPERATING 8500 HORSE-POWER OF STIRLING BOILERS

2280 HORSE-POWER INSTALLATION OF STIRLING BOILERS FOR THE HOLLY SUGAR COMPANY, HUNTINGTON BEACH, CAL.

EXTERIOR OF FIRESTONE TIRE AND RUBBER COMPANY PLANT, AKRON, OHIO

6220 HORSE-POWER INSTALLATION OF STIRLING BOILERS FOR THE SOUTHERN CALIFORNIA EDISON CO., LONG BEACH, CAL.

YABUCOA SUGAR COMPANY, SAN JUAN, PORTO RICO, OPERATING 1770 HORSE-POWER OF STIRLING BOILERS

3000 HORSE-POWER INSTALLATION OF STIRLING BOILERS FOR THE
LEHIGH PORTLAND CEMENT COMPANY, ALLENTOWN, PA.

GENERAL VIEW OF THE LAKE SHORE STATION OF THE CLEVELAND ELECTRIC ILLUMINATING CO., CLEVELAND, OHIO

PORTION OF 14000 HORSE-POWER INSTALLATION OF STIRLING BOILERS AT THE LAKE SHORE STATION OF THE CLEVELAND ELECTRIC ILLUMINATING COMPANY, CLEVELAND, OHIO

GENERAL VIEW OF GERMAN-AMERICAN PORTLAND CEMENT COMPANY PLANT, LA SALLE, ILL.

2800 HORSE-POWER INSTALLATION OF STIRLING BOILERS FOR THE
GERMAN-AMERICAN PORTLAND CEMENT COMPANY

1800 HORSE-POWER INSTALLATION OF STIRLING BOILERS FOR THE BROWN AND SHARPE MANUFACTURING COMPANY PROVIDENCE, R. I.

11000 HORSE-POWER INSTALLATION OF STIRLING BOILERS FOR THE B. F. GOODRICH RUBBER COMPANY, AKRON, OHIO

PORTION OF 19000 HORSE-POWER INSTALLATION OF STIRLING BOILERS FOR THE PITTSBURGH STEEL COMPANY, MONESSON, PA.

1900 HORSE-POWER INSTALLATION OF STIRLING BOILERS FOR THE CURTIS PUBLISHING COMPANY, PHILADELPHIA, PA.
FRANK C. ROBERTS & CO., ENGINEERS

8908967350З

B89089673503A

This book may be kept
FOURTEEN DAYS
A fine of TWO CENTS will be charged for each day the book is kept overtime.

NOV 21 '58			

Printed in the USA
CPSIA information can be obtained
at www.ICGtesting.com
LVHW020857170124
769040LV00006B/267